SIMPLE DEVICES

THE LEVER

Patricia Armentrout

The Rourke Press, Inc.
Vero Beach, Florida 32964

Patricia Armentrout specializes in nonfiction writing and has had several book series published for primary schools. She resides in Cincinnati with her husband and two children.

PHOTO CREDITS:
© Armentrout: Cover, pages 4, 7, 10, 16, 18, 19, 21; © East Coast Studios: pages 6, 15; © James P. Rowan: pages 12, 13; © Dario Perla/Intl Stock: page 9; © Steve Meyers/Intl Stock: page 22

EDITORIAL SERVICES:
Penworthy Learning Systems

Library of Congress Cataloging-in-Publication Data

Armentrout, Patricia, 1960-
    The lever / Patricia Armentrout.
        p. cm. — (Simple Devices)
    Includes index
    Summary: Text and pictures introduce the lever, a simple device used to change force and motion.
        ISBN 1-57103-177-4
    1. Levers—Juvenile literature. 2. Lifting and carrying—
Juvenile literature. [1. levers.]
I. Title II. Series: Armentrout, Patricia, 1960- Simple Devices.
TJ147.A763 1997
621.8'11—dc21   97–15149
                                                 CIP
                                                 AC

Printed in the USA

TABLE OF CONTENTS

DEVICES

If you look around, you can see many **devices** (deh VYS ez). Cars take us around town; washing devices clean our clothes; and blenders prepare our favorite frozen drink. These devices are complex—made up of many parts.

Many complex devices need fuel or electricity to run, but long ago people provided all the fuel—muscle power—to accomplish tasks. Tasks like pushing, pulling, and lifting heavy objects were done with human power until simple devices were invented.

A hammer is a type of lever.

MECHANICAL ADVANTAGE

Devices let us do tasks with less effort than muscle power alone. A simple device makes work easier by its **mechanical advantage** (mi KAN eh kul ad VAN tij). The screw, pulley, wheel, wedge, inclined plane, and **lever** (LEV er) are simple devices. All of them have a mechanical advantage.

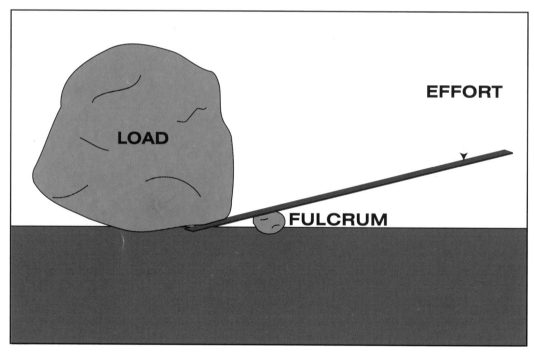

A first-class lever has a fulcrum between the effort and the load.

It is possible to lift a car using the right kind of lever.

The mechanical advantage of a lever is in three things. The first is the load. The load is what is to be lifted or moved. The second is the effort. The effort is the force used to move the load. The third is the **fulcrum** (FULL krum). The fulcrum is the support or balance, sometimes called the pivot point.

THE LEVER

Levers are used to change force and motion. Let's use a playground seesaw to see how a lever works.

The center support bar is the fulcrum, which does not move. The load is a friend sitting on one end. The effort is you at the other end trying to lift your friend off the ground.

When you push down on your seat (the force), you can lift your friend (the load), while the support bar works as the pivot point. It takes less effort to lift your friend this way than it would to simply pick up the person!

A playground see saw is a first-class lever.

FIRST-CLASS LEVERS

The seesaw works using the principle of a first-class lever. A first-class lever has the fulcrum in the middle, between the load and the effort. A hand-held can opener is also a first-class lever.

When you lift the handle of the opener (effort), the pointed end of the can opener pierces the can (load) while the notch grips the edge of the can. The fulcrum, between the effort and the load, is the can edge.

The ring-pull on a soda can is a type of lever.

SECOND-CLASS LEVERS

A wheelbarrow is an example of a second-class lever. Second-class levers have the load in the middle. The load is between the fulcrum and the effort.

Can you find the effort, fulcrum, and load in this lever?

A wheelbarrow is a second-class lever that has the load in the middle.

Imagine a wheelbarrow filled with dirt. The dirt is the load. On one side of the load is you lifting the handles, applying the effort. On the other side of the load is the wheel—the fulcrum. The wheel acts as pivot while the effort lifts the load.

THIRD-CLASS LEVERS

An example of a third-class lever is a drawbridge on a castle. It too has the three parts of a lever, but the effort is in the middle. The effort is between the fulcrum and the load.

The bridge itself is the load. The hinge that attaches the bridge to the castle wall is the fulcrum. In between are chains (effort) used to draw up the bridge.

A castle drawbridge is an example of a third-class lever.

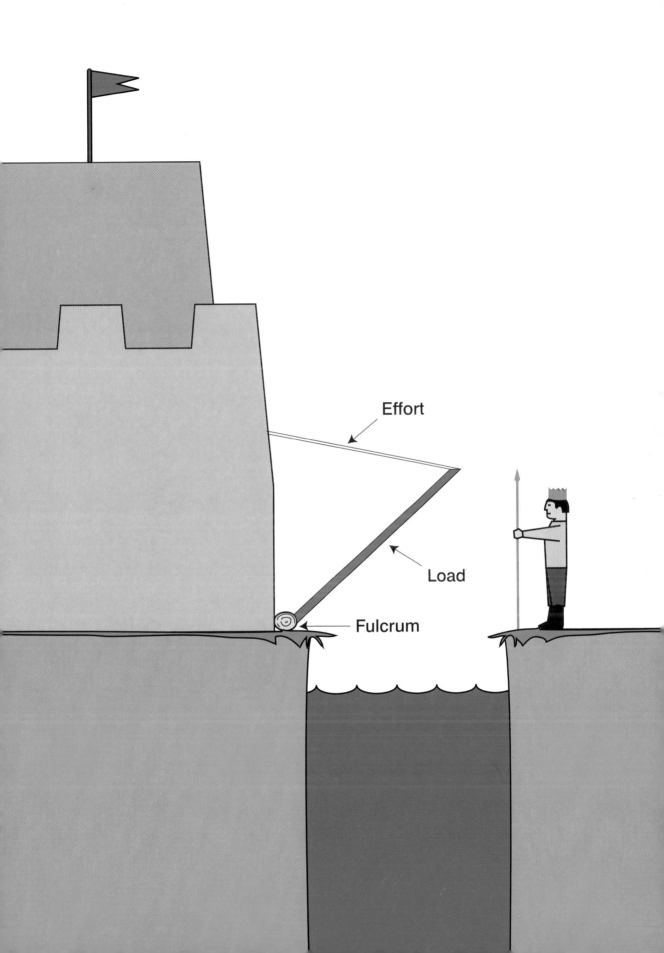

HOUSEHOLD LEVERS

Try this household task and figure out what type of lever you're using: Sweep the kitchen floor. Yes, a broom is a lever!

To sweep dirt into a pile you probably used two hands on the broom handle, one hand above the other.

Did you decide that the broom is a third-class lever? Your upper hand (the fulcrum) holds the broom while your lower hand (the effort) sweeps the dirt (the load) across the floor. The effort is between the fulcrum and the load.

The effort in the middle makes this broom a third-class lever.

DOUBLE LEVERS

Now that you understand what a lever can do—let's put two levers together. Two levers joined together are called double levers.

Scissors are two levers joined by a fulcrum.

Your hand provides the effort when you use a hole punch.

You can find double levers around the house. Scissors, pliers, and nutcrackers are all double levers. Each half of the tool is a lever, joined by the fulcrum. When using any of these double levers your hand provides the effort, or force.

SIMPLE DEVICES WORKING TOGETHER

Complex devices, like cars, consist of many parts—some parts being simple devices. Simple devices work together to make a new device. Take the corkscrew for example. The screw itself is a simple device. When joined with a lever, they work together to remove a cork from a bottle.

Can you think of other objects used as levers? Remember a lever has three parts—a load, an effort, and a fulcrum that doesn't move.

A firefighter's wrench is a lever that is used to open fire hydrants.

GLOSSARY

device (deh VYS) — an object, such as a lever, pulley, or inclined plane, used to do one or more simple tasks

fulcrum (FULL krum) — the support on which a lever turns

lever (LEV er) — a simple device used to change force and motion

mechanical advantage
(mi KAN eh kul   ad VAN tij) — what you gain when a device allows you to do work with less effort

The fulcrum is where the ore is attached to the boat.

INDEX